AJS & MATCHLESS POSTWAR TWINS

1948 - 1969

Roy Bacon

NITON PUBLISHING

First published in the United Kingdom by:
Niton Publishing
PO Box 3 . Ventnor . Isle of Wight PO38 2AS

Acknowledgements
The author would like to thank those who
helped this book by supplying the photo-
graphs. Most came from the EMAP archives
or *Motor Cycle News* by courtesy of the
Editor, Malcolm Gough, with others
from the Mike Woollett archive.
My thanks to all.

Filmset and printed by Crossprint
Newport . Isle of Wight

ISBN 0 9514204 7 X
A CIP catalogue record for this book is
available from the British Library

Front Cover: The 1956 Matchless G11 twin as first advertised in the show
issue of *Motor Cycling* late the previous year.

Back Cover: A 1960 *Motor Cycle* show issue showing the AJS twin
winning the 1960 Thruxton production race.

D.J.P. Wilkins waits to start in the 1951 Clubman's TT with his Matchless
G9. Open pipes and no electrics were allowed at that time.

Contents

Introduction

AJS and Matchless became one firm, based at Plumstead in South London, in 1931, when the former went into liquidation and was bought by the latter. In the late 1930s, they acquired the Sunbeam firm and registered the trio as Associated Motor Cycles or AMC. They soon sold the new acquisition to BSA, though not before learning the secrets of Sunbeam's excellent paint finish.

During the war, AMC concentrated production on one model, the G3L Matchless, and post-war built a pair of staid singles in two capacities and under each marque. They differed little, most parts being common to both makes, and often both engine sizes. This policy of standardisation was to increase as the post-war years rolled by.

It took them until late 1948 to put their vertical twin on the market, as the first few years after 1945 had to be spent with the accent on the production of existing single cylinder designs and exports. Materials and parts were in short supply, while war-time restrictions hung on, often far too long. There was little time to carry out much improvement, which had to wait while the essentials of earning a living took priority.

Once this need eased a little, it was possible to plan new models, and AMC combined the launch of the new twin with new singles featuring rear suspension. The twins were always to have this, so many of the cycle parts were common, which helped development.

The first twins were the Model 20 AJS and G9 Matchless, which were built through to 1961. For 1959, the standard versions were joined by CS and CSR sports twins, but before then a 593 cc twin was introduced as the Model 30 and G11. There were sports versions of this for 1958, but for 1959 the 593 cc models were replaced by the 646 cc 31 and G12. These came in with standard and sports versions, which continued up to 1966.

Before then, in 1965, AMC turned to using 745 cc Norton twin engines to produce the 33 and G15 models, of which the AJS versions were dropped at the end of 1967, and the Matchless models in 1969. Prior to them, there had been two batches of 750 cc Matchless G15 or G15/45 twins, which had been built for export using the AMC engine and a variety of the sports cycle parts. These were not the only special twins built for export, as around 1954 there were some 550 cc twins made and listed as the Matchless G9B or G10.

The AMC engine was unique among the post-war British vertical twins in having a centre main bearing, and it was generally a good, reliable and well-liked unit. There were problems, especially in later years, but the firm had an excellent reputation for the quality of its products, and they were much admired and sought after. The advertising was always marque orientated, even if all the copy came from the same office, and this had its effect. In the clubrooms of the 1950s, riders would debate the merits of AJS and Matchless with some fervour, even though they knew perfectly well that most of the parts were common and all went down the same production line.

Would that AMC had kept that loyalty in the mid-1960s, but it was not to be.

John Morris with his 1961 Matchless G12CSR after winning the Motor Cyclist of the Year Trophy for 1962.

Saddle or dualseat - 20 or G9

The AMC twins were announced late in 1948 and shown at Earls Court but, tantalisingly, they were for export only at first. Each had its own style, which was quite a feat when the number of common parts was considered, and this remained so for a long time.

All AMC twins had telescopic front and pivoted-fork rear suspension, a separate gearbox and single-leading-shoe drum brakes. Many parts were common to both makes and also to the single-cylinder range, the gearbox, front forks and wheels being three such assemblies. The result, when the classic revival of the 1980s came, was to make many parts easy to locate as spares, and for most to be well developed and reliable.

The heart of any machine is the engine, and for their twin, AMC went down the well-trodden route pioneered by Edward Turner, in 1937, with his Triumph Speed Twin. Once one accepted the need to use a standard twin-cylinder magneto and a single carburettor, there was no option but the 360-degree layout with parallel cylinders. All the major British firms went the same way,

Start of the line for AJS was this 1949 model 20 with candlestick rear units, saddle and separate pillion seat .

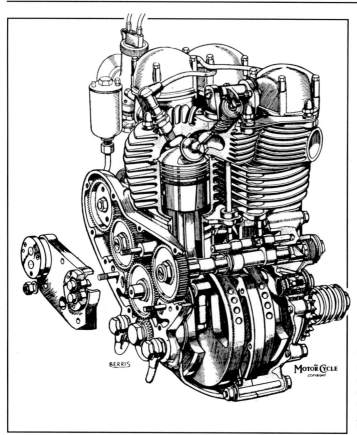

BERRIS

MOTOR CYCLE
COPYRIGHT

The AMC engine as
used by both AJS
and Matchless in
498, 593 and 646
cc capacities with
just a change of
timing cover.

and all, in time, paid the penalty of
excess vibration when the engine
capacity and speed became too
much.

AMC had none of this trouble at
first, for their overhead-valve engine was of 498 cc capacity and its
dimensions were 66 x 72.8 mm, so it
was not too much of a long-stroke. It
also had its third, central, main
bearing and differed from most in
that the cylinders and heads were
separate, so only the inlet manifold
and a tie plate connected the two

top halves.

The bottom half was constructed
in the fashion of the time and based
on an aluminium crankcase, the
castings of which were joined on the
vertical centre-line of the engine.
They were aligned to each other by
the central panel, which carried the
third main and was fixed to the left-
hand, drive-side, crankcase half by
six studs and nuts.

The centre bearing was plain with
split shells, each of which had a
flange on each side to take the end

This is a 1951 Matchless G9 but that marque was fitted with a dualseat and the lovely megaphone shaped silencers from the start.

thrust and locate the crankshaft. The shells and shaft were held in place by a cap that was fitted on two studs from below. The bearing itself acted as the feed point for the big-end oil supply which, thus, was equally distributed and not all from one side, as in most other twins of that time.

The bearing housing helped to take some of the crankshaft load and prevented it whipping, but its location was none too rigid, so the total effect was perhaps less than expected. Some firms preferred the lower friction of just two mains and allowed the crankshaft to flex with-

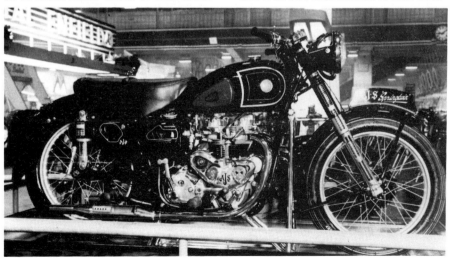

Sectioned 1953 AJS model 20 on show at Earls Court late the previous year. Jampots have replaced the candlesticks.

A 1953 Matchless G9 during tests carried out when 80 octane petrol replaced the poor Pool used from the outbreak of war in 1939.

out any real restraint. AMC preferred their method and stuck with it.

The crankshaft was in one piece and unusual in being cast in a high-grade alloy-iron, rather than being the usual steel forging, so it was not as flexible as the latter. It was formed with bob-weights as the outer wheels, but as full circles for the inners, although these were recessed to match the outers and help balance the engine. Oil holes were drilled from the centre shaft to each crank-pin, and each was sealed with a small screw. Thus, the ends of the holes acted as sludge traps, and removal of the screws enabled them to be cleaned.

Because the mainshafts were cast-iron, they had to be larger than normal, which dictated the use of big main bearings at each end. These were roller races with a 1-3/8 in. bore and 3 in. outer diameter, so they were truly massive. A washer went between each bearing and the web cheek. The same bearings were

The Matchless G9 for 1952 when there was less chrome plating due to shortages but the finish remained to the high AMC standard.

to serve all AMC twin engines, regardless of size or year.

With separate top halves, it was essential to maintain a rigid crankcase, so the top of this was not cast as an opening, as was usual with a twin having a cylinder block. Instead, each upper face was accu-

rately bored to take the very deep spigot of the cast-iron cylinder that sat on a machined face.

The two cylinders were well spaced for good cooling and spigoted into each cylinder head. Each head and barrel assembly was retained by four long studs with dome

First year with a full-width front hub was 1954 with it seen here on a Matchless G9.

nuts, while suitable drilled holes gave passage for the pushrods and for oil feeds. Each barrel spigot was also machined with two grooves, the upper of which was the rocker box oil feed gallery, and the lower

A compression ratio of 7:1 was used and the pistons domed with valve cut-aways. A hollow gudgeon pin, retained by circlips, attached each piston to its connecting rod, running directly in the small-end.

The twins were quite capable of pulling a sidecar along as this 1955 AJS model 20 with full-width hubs front and rear shows.

the feed for the camshaft.

The pistons were split-skirt and wire-wound to minimise slap and promote quiet running. The latter aspect was already in use in AMC singles, each piston being grooved just below the scraper ring to enable a high-tensile steel wire to be wound on to it. Once in place and secured, the wire was ground to size, flush with the skirt.

The rods were forged in light-alloy, with shells at the big-end, and the caps were retained by studs anchored in steel trunnions.

The cylinder head for each barrel was an aluminium casting and was extended upwards to form two supports for each rocker spindle. A gasket went between the head and barrel, with another at the crankcase top face. The valve seats were

By 1957, when this AJS model 20 was built, the twins had a revised frame, new oil tank shape and the AMC gearbox.

shrunk-in and the guides pressed in to a circlip, which located them. The valves were controlled by duplex springs, which sat on seats located on their guides, and each was retained by a collar and split collets.

Each rocker fitted between the spindle supports, with thrust washers and a spring to remove sideplay, and each had two bushes pressed into it. The rockers oscillated on eccentric spindles, which

This is the larger 593 cc AJS model 30 which shared cycle parts with the smaller twin and the Matchless models.

The 1958 Matchless G9 still retained the megaphone silencers but had lost the pilot side lamps used for a short time.

could be turned to set the valve gaps. Thus, there were no adjusters in the rockers, only ball ends on the inner arms to match the pushrod cups. Small bolts in the supports locked the spindles by clamping their heads once the valve gaps had been set. Each complete assembly of rocker, supports and spindle had a polished-alloy, domed cover held by four screws to enclose the valve gear fully.

The pushrods were assemblies, comprising two hardened end pieces

Engine unit of the 1958 Matchless G11CS sports twin which offered more performance and skimpier mudguards.

The 1958 Matchless G11 fitted with tank side panels and with a lighter finish for the mudguards and tank.

pressed on to an alloy centre, and were positioned front and rear, at the inner corners of the barrels, but outboard of the fixing stud holes. Thus, there were two camshafts that sat high up in the crankcase to fore and aft of the crankshaft, each being supported by three bushes. One of these was plain and went in the drive-side case, but the other two had flanges and were fitted to the timing side, at each end of the case-wall hole.

Above each cam sat a lever follower, with a hollow in its top surface for the pushrod to sit in. A rod ran across between the case halves for each follower pair to pivot on, and a distance piece kept them in the right place. The followers were to become a wear point in the engine and a bind to change, as this meant splitting the crankcase.

The camshafts were gear driven, a pinion being fitted on the right-hand end of the crankshaft and an intermediate gear above it. This had the same number of teeth as the camshaft gears and was bushed to run on a fixed pin pressed into the crankcase wall. The gear train ran on to both front and rear to drive the

Matchless G12 of 1959 attached to a Jet 80 sidecar and on show with others from the range.

electrics, and the small front pinion was fixed to a Lucas dynamo. This was held to the front of the crankcase by a strap, and a hole in the back of the timing chest allowed the pinion to pass through to mesh with the camshaft gear. Thus, the dynamo could be readily removed. At the rear, the gear train ran on to a manual-advance Lucas K2F magneto, which was flange-mounted to the back of the timing chest.

Both ends of both camshafts were used to drive items involved with the dry-sump lubrication system. This was quite complex and used two gear pumps for feed and scavenge, which were mounted on a single plate that was fitted into the timing chest and also supported the outer end of the intermediate gear spindle. The pumps were driven from the camshaft ends and were physically interchangeable, but correctly the exhaust was coupled to the feed pump and the inlet to the larger scavenge unit. Reversal would immediately cause oil to accumulate in the crankcase.

The oil system's external tank was connected by two pipes, with banjo fittings, to the right-hand crankcase side, just below the timing chest. The feed pump had a bleed across the mounting plate to keep the return pump primed and ready to function. It was fed via a mesh filter, and its output went to a felt filter located in a crankcase chamber. This had a relief valve at one end, which was additional to the main valve located next to the feed connection.

From the filter, the oil went through an anti-drain valve to a junction. The main supply from this went to the centre main and then on to the big-end bearings but some was diverted to a distributor bush that was driven by the left-hand end of the exhaust camshaft. This sent oil alternately to the grooves in the barrel spigots which, in turn, supplied the rocker spindles via meter-

ing plugs and the camshaft tunnels via internal drillings. In this way, the top end was kept well oiled and the cams had a supply to dip into as they rotated.

The left-hand end of the inlet camshaft was given a timed breather valve to drive, and this was connected to an external fitting and a pipe that ran back to the oil tank. Both the items at the left-hand ends of the camshafts were accessible by removing screwed plugs in the crankcase, while a pressure gauge could readily be connected in place of the filter relief valve for a check.

The whole of the timing chest,

The G9CSR of 1959 with siamesed exhaust pipes, polished alloy mudguards and shorter dualseat.

A 1959 AJS model 31 out on the road. This was the first year for the 646 cc twin and this standard one has alternator electrics.

with the gear train and oil pumps, was enclosed by a single cover that was held by ten screws. This was one of the few items to differ between the marques, the Matchless version being thicker than the AJS one. It was smoother with just a recess over the area around the crankshaft end, in which a 'flying-M'

symbol was formed. Being thinner, the AJS cover had two bulges to clear the oil pumps, the marque letters being added between them.

An Amal type 6 carburettor, of 1 in. bore, provided the mixture and was itself supplied from a separate float chamber. The Amal was fitted to an aluminium manifold, which was bolted to the two cylinder heads, and no air filter was offered, even as an option. For the exhaust, there were down-pipes and silencers on each side, the makes differing in the latter. The AJS used a conventional tubular silencer, but the Matchless had a megaphone, which did its job well and gave the machine an excellent line.

was sealed at the join by a rubber band that was held in place by a metal one. This was clamped together by a screw at the rear of the case. The arrangement was much as for the singles, except that it did not enclose the dynamo drive as well, and was never easy to keep oil-tight. The rear chain ran from a sprocket inboard of the clutch and

The 1959 AJS model 20 standard model with alternator inside its cast-alloy chaincase. Tank in optional two-tone finish.

Primary drive was by a single-strand chain from an engine sprocket incorporating a two-lobe shock absorber. It drove a multi-plate clutch, which was clamped by five compression springs and mounted on the mainshaft of the Burman gearbox. This was a typical British design, with four speeds and footchange, the pedal being on the right, as was the kickstarter.

The drive was enclosed by a two-part, pressed-steel chaincase, which

was protected by a guard over the top run.

The engine and gearbox went into a built-up frame with pivoted-fork rear and telescopic front suspension. The frame was of traditional, brazed tube-and-lug construction, and the front section comprised the headstock plus top, down and seat tubes. From the base of the downtube, a rail ran back on each side of the engine, passed the rear fork pivot, and terminated with the pillion footrest and silencer support.

A substantial aluminium bridge casting went between the two rails and the seat tube, and it had the rear fork pivot pin pressed into it. The fork legs had bronze bushes at their forward ends to pivot on the hollow pin, but dismantling and assembly required the pin to be moved. Lubrication of the bearings was done by filling the pin with oil, capping its ends and allowing the oil to seep on to the working surfaces, felt seals being fitted to keep it in place. The subframe was formed from two side loops with a single cross tube, and it was bolted to the main frame and the bridge.

The rear suspension units were attached to upper supports that were welded to the frame loops and had rubber-bushed clevis fittings at top and bottom. The units were made by AMC and were quite slim so, in time, became known as 'candlesticks'. Each had a single spring, with no provision for adjustment to cope with the added load of a passenger, and included hydraulic damping. It was feasible to dismantle them for repair, but they were to prove rather inadequate in operation and very sensitive to temperature and the volume of oil in them.

The front forks were the AMC Teledraulics, which had first been used in 1941 on the wartime single. They had light-alloy lower legs with

The AJS model 31CSR fitted with siamesed exhaust pipes and offering considerable performance.

Vic Willoughby out on a 1959 Matchless G12CSR which he rode far and fast as was his habit.

the springs above, hydraulic damping and top covers in two parts, one to conceal the springs and the other to support the headlamp shell. A steering damper was fitted under the lower crown and was controlled by a handwheel at the top of the forks.

Plates held the engine and gearbox in the frame and were provided with a means of moving the latter to adjust the primary chain tension. The oil tank was fitted above the gearbox on the right, and was matched by the battery and its carrier on the left. A toolbox went into the corner of the subframe on each side, and centre, front and prop stands were fitted, the last being a fly-back type common to AMC and not much liked by riders.

The front stand was also the rear

stay of the front mudguard, which was held by a front stay and centre bridge as well. The guard was sporting, so without a valance, but the rear one retained this feature, as much for its own strength and rigidity as for keeping road dirt at bay. The tail of the rear guard could be easily detached to allow the wheel to roll out when required and an optional rear carrier could be added if desired.

Both wheels had hubs that ran on taper-roller bearings, and the front was unusual in that the wheel spindle was the inner race, so it came with the two roller cages as an assembly. In the past, this had applied to the rear as well but, for the twins, two separate roller races were used.

The front hub had a large spoke

Matchless G12 standard model of 1959 with alternator electrics and optional finish for petrol tank, mudguards and centre panels.

The 1960 AJS 31 standard model in the new duplex frame with alternator and optional tank finish.

flange on the left, and the 7 in. brake drum was fixed to this by a row of ten 1BA bolts. The backplate for the single-leading-shoe brake was anchored by a stay, which was fixed to the inside of the fork leg and provided a threaded hole for the operating cable adjuster.

At the rear, both spoke flanges were the same size, and the 7 in. brake drum with integral rear sprocket was bolted to the hub. The backplate located to the rear fork leg, and the single-leading-shoe brake was rod actuated by a pedal on the left. The rod included the usual knurled nut for adjustment, but both brakes had an additional internal method of adjustment. This featured a thrust pin in each brake shoe to take the wear of the cam turning against it. By fitting one or more washers behind the pin's head, it was possible to move the brake shoe and its lining closer to the drum.

The rear wheel spindle carried a cam on each end, and these were keyed in place so that they could control the wheel alignment while the rear chain tension was set. Both wheels were spoked into 19 in. steel rims, and the front was shod with a 3.25 in. section tyre, while the rear had a 3.50 in. one. The rear wheel drove the speedometer, which read up to 120 mph or 180 kph as applicable and was mounted on the top fork crown.

The electrics were standard Lucas for the time and were based on a six volt system, the dynamo output being controlled by a unit fixed to the battery carrier. The horn went under the front of the tank, and the headlight had a small panel set in it to carry the ammeter and light switch. On the handlebars were a dip-switch, horn button and magneto cut-out, along with the twistgrip and levers. Both of the last were of the combined type, with brake and air on the right, and clutch and ignition to the left.

The finishing touch to any machine is the petrol tank, and in this, and the seating, the two marques differed. The AJS tank was the larger at four Imperial gallons, while the Matchless tank, although of similar lines, only held three. For seating, the AJS had a saddle and a pillion pad on the rear mudguard, but the Matchless was supplied with a Dunlopillo dualseat.

The finish of both machines was mainly in black, except for the tank and wheel rims. Both tanks were chrome-plated, but the AJS top and side panels were black, while the Matchless ones were in red. Both were lined in silver, but the AJS had a blue pinstripe as well, and it used transfers for its badges, while the Matchless had its letter 'M' badge. The wheel rims were chrome-plated with gold-lined, black centres for the AJS, and silver-lined, red ones for the Matchless.

Both machines carried a small round badge set in the side of the drive-side crankcase, near the top, and were given names as well as model numbers. Thus, the AJS Model 20 was also the 'Springtwin' and the G9 Matchless the 'Super Clubman'. Of the two, the G9 was perhaps the more handsome, thanks to the red tank finish, dualseat and megaphone silencers, but this was academic to home buyers, who could look, but not touch for some while. The basic price, without UK purchase tax, was £165 for the AJS, but £167 for the G9.

Early developments

A few twins reached their home market in 1949, where they soon found eager buyers and quickly gained a good reputation. They went well, and handled and ran in a manner that made for easy riding. The brakes were not really up to the performance and needed to be larger, but this was not recorded in an official road test. At that time, AMC refused to offer machines to the press as other firms did, and this continued up to the late 1950s, except for brief outings with competition mounts.

With demand high, there was little call for modifications, so 1950 simply brought ribbed mudguards and a new silencer, with offset inlet and exit tubes, for the AJS alone. There was more for 1951, and in the engine the centre main bearing

Matchless G12CS of 1960 with small tank, detachable headlamp and sports fittings for off-road or trail use.

The sports 1960 AJS model 31CSR with magneto and dynamo from the past but new duplex frame for the future.

became a pair of shells, each with two separate flanges. This was conventional car practice and much easier to produce. In addition, the oil deflector, which had helped to direct oil to the cams, was no longer fitted.

On the cycle side, the most obvious change for 1951 was to the rear suspension units, which were made much larger in diameter. They were immediately called 'jampots' and the name stuck to become standard terminology in motorcycling and the title for the UK owners' club magazine. The new design held more oil

Standard model 1960 Matchless G12 with optional finish for tank and mudguards to make a stylish and fast tourer.

and worked at lower internal pressures than the 'candlestick', so it was a real improvement.

Among the details altered that year was the recessing of the front fork drain plugs to prevent them being wiped off when parking by a high kerb. The centre stand legs

the left-hand crankshaft end and had a simple, spring-loaded flap-valve. This was connected via a mainshaft drilling to the crankcase and could be fitted to early engines if this essential 1/4 in. hole was run through. The new device just breathed into the chaincase, which was vented to

The standard Matchless G9 of 1961 in its standard and rather sombre black finish. Options allowed it to be brighter.

were lengthened, and the addition of an air filter option meant that the frame was modified to accommodate it and its connection pipe. For the Matchless, the dualseat cover material was changed to vynide, and the AJS tank was fitted with oval badges in place of the transfers.

The main change for 1952 was to a Burman B52 gearbox, which had a number of improvements over the older type. It meant that the clutch adjustment had to become a screw in the pressure plate, so an access cap was added to the outer chaincase. For the engine, there was a new breather, which was fitted into

the atmosphere.

For the 1952 cycle parts, there was a light-alloy front brake backplate and an underslung pilot light beneath the headlamp. The finish was amended to take account of the nickel shortage caused by the demands of the Korean war, so the wheel rims had an aluminium-effect finish, called Argenized by AMC. The petrol tanks were both black, but the lining was gold for AJS and silver with a red pinstripe for Matchless. There were new light-alloy, die-cast badges for both, the AJS version being formed as the marque name, and the Matchless one as a disc with

Jim Goldsmith with sponsor Norman Reeves and the AJS sports twin CS model he rode

successfully in the late 1950s.

the 'flying-M' on a red background.

The AJS finally got its dualseat for 1953, when the engines had modified cam followers and rocker covers that were only held in place by two bolts each. The front brake cam lever was turned so that it pointed forward instead of aft, a plastic rear lamp went in place of the earlier metal-bodied one, and a

models continued in the 1949 style, with a chrome-plated tank featuring lined panels, as the restrictions of the past were no more. Once more the wheel rims were plated, painted and lined, but the front hub was left natural. This was because it was changed to a full-width, light-alloy type with central cooling fins and shrunk-in liner. The new hub ac-

AJS standard model 20 of 1961 with the optional finish. A nice motorcycle in its final year.

steering-lock bar was available. The finish stayed as it was in 1951 for the home market, but export models went back to the chrome-plating of 1949 for the wheel rims and tank, although they had the AJS oval and Matchless round tank badges.

The two marques shared a common petrol tank for the first time in 1954, when both changed to round plastic badges. The finish for all

cepted straight spokes, but its brake was no larger than before, so an opportunity had been missed.

The mudguards were flared, and access to the clutch was improved by making the chaincase dome that fitted over it a separate part, which was an improvement over the 1952 access cap. The underslung pilot lamp went, but was replaced by two small lights positioned one on each

The AJS 31 de luxe model of 1961 which retained its old-style electrics for its last year in the lists.

side of the headlamp shell, although they were no more easily seen.

The style of the new front hub was nothing exceptional, so it was altered for 1955. In its new form, it was narrower and the fins were formed in a barrel profile, which looked much better. The rear hub copied the front to become a full-width, light-alloy assembly, while both were given the final machining of the brake drum after the wheel had been built-up and trued - an excellent practice.

Other changes for 1955 included larger-diameter forks and modified jampots to improve the suspension. The headlamp shell was made deeper so that it could accommodate the speedometer, with the ammeter and light switch ahead of this, while the front mudguard no longer had a forward stay. The supports for the pillion footrests became pressed-steel lugs, and there was a new silencer for the AJS model. Both twins changed to a Monobloc carburettor, which remained at the 1 in. size, and the finish was as before.

The AMC twins were now well established and popular, but the firm, as with others, was pressured to increase the engine capacity for more power and speed. A partial response to this had been made in 1954 with the Matchless export G9B or G10 model, which was opened up to 550 cc. This was, however, only an interim step, and by the end of 1955, the time had come for a larger engine and a new frame.

Enter the 600

The larger AMC twins were announced for the 1956 season as the AJS Model 30 and Matchless G11. Their 593 cc capacity came from boring out the cylinders to 72 mm, while retaining the 72.8 mm stroke, and the crankshafts differed from those of the smaller engine due to the heavier pistons of the larger engine.

Little else was altered, other than the barrels and pistons to suit the larger bore, the exhaust pipes (but not the silencers), the sidecar gearing (but not the solo), the transfers and the tank finish. The carburettor remained the same 1 in. Monobloc, but the compression ratio was 7.5:1,

while that of the 498 cc engine went up to 8.0:1. The rest of the engine was common to both sizes, as was the rest of the machine, much of it in revised form.

The gearbox remained the Burman B52 and the primary transmission and chaincase were as before, with the big access dome in the outer cover to enable the clutch to be serviced. The front forks and rear jampots stayed as they were, as did the wheels with their full-width alloy hubs, but the frame itself was revised. The main change was that the seat tube was vertical and brazed into a massive, malleable-iron lug, which incorporated the rear fork

The 1961 AJS 31CSR sports twin with siamesed exhausts, alloy mudguards and short dualseat.

A 1961 standard model Matchless G12 showing its typical marque lines which remained much the same for many years.

pivot. In other respects, the frame was as before, with front section, rear loop assembly and two side rails that were bolted in place. The disappearance of the alloy bridge meant that servicing the fork pivot was more difficult, as the whole front frame had to be handled under the press.

The area under the dualseat was restyled for this frame, so the oil tank became long and thin to fit

The Matchless G12CS of 1961 with its small petrol tank and other off-road features.

For this 1962 AJS model 31 Swift there were new tank badges but little else changed.

along the right-hand side into the subframe loop. A shield, or cover, was attached to the outer side of the tank by two screws and formed to leave a gap for cooling air between itself and the tank. The tank was balanced on the left by a combined toolbox and battery carrier of matching shape, and the tall air filter went between the two. A cover was fitted over the gearbox, while the horn was moved under the seat, and the front mudguard stay no longer doubled as a stand. The front brake cam was moved to the top of the backplate, but its lever continued to point forward although it became curved to clear the fork leg.

The finish for the Model 20 and G9 remained as it was, but the larger twins had their tank sides chrome-plated with the top in black for the

The 1962 Matchless G12 Majestic was also fitted with new tank badges which became known as 'knee-knockers'.

Model 30, and red for the G11. The rims continued to be finished to match, while the rest of the painted parts were in black.

In May 1956, the whole of the AMC range changed over to a new gearbox, which also appeared on all Norton models, they having become part of the AMC empire. It was based heavily on the original Norton gearbox of 1935, which had been taken over by them from Sturmey-Archer, improved in minor details, and manufactured by Burman. The position of the positive-stop mechanism had been moved in 1949 to suit the Norton twin, and the bulk of the new box was as this type. The clutch mechanism and positive-stop details were altered so that the end cover was smaller, but the shafts, gears, camplate and selectors were effectively unchanged. It meant new lessons for AMC dealers and owners, and the box was always known as the AMC type, regardless of the marque it was fitted to.

The AMC gearbox included a

A 1962 AJS model 31CSR Hurricane, which had been used for travelling marshal duties at the TT, on a quick trip to Germany.

clutch with a shock absorber built into its centre hub, so there was no longer any need for the engine shock absorber which was dropped for 1957. For that year, and with this out of the way, the dome in the outer chaincase was reduced in size so that it just cleared the crankshaft end. The larger engines were given a slight increase in Monobloc size to 1-1/16 in., and the oil tank cover and toolbox lid gained small ribs. The jampots were finally laid to rest, for

damper technology had moved on to become a specialised function. AMC bowed to this and fitted Girling units. While these had conventional top mountings, which meant a frame modification for AMC, they retained the clevis lower ends, so they were unique to the marque.

The petrol tank was the one area to have a revised finish for 1957, and all models changed to having painted tanks with separate, chrome-plated side panels. These ran the length of

The short-lived headlight cowl that was an option for 1962 only and here seen fitted to a Matchless G12CSR Monarch of that year.

the tank and were held in place by the badge and kneegrips on each side, with a coloured beading round the edge. Both 498 cc tanks were black, but with a light blue beading for the AJS, and red for the G9. On the larger models, the AJS kept the beading colour, but on a Royal blue tank, while the G11 was in red with black beading.

gasket. This, at last, kept the oil inside. The pilot lights went, being replaced by a bulb set in the main reflector, so another irritation was no more. The finish was as for 1957, but with chrome-plated wheel rims, except on the Model 30, which kept to blue centres. The export models had the oil tank cover and toolbox lid in blue for AJS, and red for Match-

A 1963 AJS 31CSR Hurricane out on the road during a magazine road test where it acquitted itself well.

The range of twins was expanded for 1958 with the addition of two sports models of 593 cc capacity. These were designated by the letters CS or CSR after the model number, such as G11CSR. All models changed to a very welcome cast alloy primary chaincase, which was well polished and used 14 screws to hold the two halves together on a

less.

The two sports twins had engines with compression ratios raised to 8.5:1 and siamesed exhaust systems feeding single silencers low down on the right. For both marques, this was a tubular type and the Matchless models had to manage without their usual megaphone shape. The engine was installed in a

Show model Matchless G12 Majestic for 1963, by when the megaphone silencers were a fond memory.

frame that was also used by the single-cylinder scrambles model, and it was similar to the road one, but not the same. Forks, hubs, oil tank and toolbox were as for the road machines, but the mudguards were polished light-alloy and of a sports section.

The CS model was built in a street scrambler mould, much as the single-cylinder machines, which were scrambles models with optional lighting. The twins followed suit with off-road tyres of 3.50 in. section at the front and 4.00 in. at the rear on 19 in. rims, while it was easy enough to fit the single's front wheel with its 21 in. rim and tyre.

For 1964 AMC adopted Norton forks and hubs for their twins and one result was this AJS model 31 Swift.

The lighting system was quickly detachable and of the older style, so the speedometer was moved back to the fork crown, while a short competition dualseat and two gallon petrol tank were fitted. The handlebars could be standard, or raised in Western style, and there were further options of a three gallon tank plus dualseat to suit it, twin low-level exhaust pipes and silenc-

blue or red oil tank cover and tool-box lid.

The CSR models were announced in January 1958 as high-speed sports machines, and they combined the sports CS engine, exhaust system, frame, seat and mudguards in a road-going form. In AMC parlance, the suffix letters stood for Competition Sprung Roadster, but instantly became 'Coffee Shop Racer', and this

The Matchless G12 Majestic of 1964, also with Norton forks and hubs, plus a siamesed exhaust system.

ers, a stop-light and pillion footrests.

The finish of the CS twins was mainly black with polished alloy mudguards, chaincase and lower fork legs. The petrol tanks were in black with gold lining and name transfer for AJS, but silver for Matchless. The optional tank colour was blue or red with chrome-plated side panels and light blue or black beading, while export models had the

name was to stay with them from then on.

They kept to the normal road tyres, so the front mudguard, curved to clear a 21 in. wheel, looked slightly wrong, but the short seat, allied to the standard tank, made up for it. The finish was brighter than for the CS, for as well as all the polished alloy items, there was chrome-plating for the fork shrouds and rear

Matchless G12 Majestic with smaller tank badge for 1965 but otherwise little changed.

unit covers together with the usual exhaust system and wheel rims. The petrol tank had chrome-plated side panels with blue or black beadings for AJS and Matchless respectively, while the main colours were Mediterranean blue for the former, and red for the latter. These were applied to the petrol tank, oil tank cover and toolbox lid.

Just how fast the new model ran was demonstrated early in the year when Vic Willoughby of *The Motor Cycle* rode a Matchless round the MIRA high-speed track for one hour. It was a cold ride, but Vic and the machine managed to cover some 103 miles in those 60 minutes, despite a touch of fuel starvation near the end.

The larger twins were a success, but most other firms had long since stretched their's out to 650 cc or more, so the pressure was on AMC to follow suit. This they did for 1959, which meant a short life for the 593 cc twins.

More models and 646 cc

The range was increased to eight models of each marque for 1959. Four still used the old 498 cc engine, but the other four had a new 646 cc unit. Each capacity came in standard, de-luxe, CS and CSR forms, and each of these in either marque. The type designations for the smaller twins remained 20 and G9, while the larger machines became the 31 and G12, and both used the suffixes as before.

The larger capacity of the new engine came from an increase in stroke to 79.3 mm while retaining the 72 mm bore of the 593 cc engine. The construction was as that of the original twin, but with longer barrels, each with one more fin. The compression ratio was 7.5:1 for the basic engine, but 8.5:1 for the sports versions. All were fitted with Monobloc carburettors of 1-1/8 in. bore.

The two basic touring and sporting model types were very much as they were in 1958, but the former was offered in standard and de-luxe forms. Both had a new 4-1/4 gallon petrol tank, also fitted to the CSR, and deeper mudguards without the centre rib. The standard models were alone in having coil ignition, controlled by an ignition switch in the centre of the lighting one, so a distributor was fitted in place of the

AJS model 31CSR for 1965 when a diamond tank badge joined the revised oil tank and toolbox shape of the year before.

magneto. A Lucas alternator went on the left-hand end of the crankshaft, with the stator located in the outer chaincase, so the crankcase of the standard models differed from the others.

Other than the electrics, the main difference between the standard and de-luxe twins lay in the finish, which was complicated by the options also available. The standard model had tion were the chrome-plated tank side panels and their beading. The second option was much the same, except for the petrol tank, which was two-tone with a chrome-plated dividing strip. The AJS tank had the tank top in blue with the lower section in light grey, while the Matchless top was Arctic white with a red lower half.

The sports twins followed the

AJS Swift model 31 for 1966, its final year but still with the line it began with in 1949.

the basic finish of black with the tank lined in gold for AJS, and silver for Matchless. The de-luxe model was the same except, for the fitting of chrome-plated tank side panels, which were an option for the standard model.

The other options applied to both versions, and the first had the petrol tank, oil tank and cover, toolbox complete with lid, and mudguards in blue for AJS, and Arctic white for Matchless. Included with this op-

lines of the previous year, with the large petrol tank for the CSR and the two gallon one for the CS. The latter also had a 3.00 x 21 in. front tyre and a 4.00 x 19 in. rear, while the CSR adopted the headlamp shell from the de-luxe model with the speedometer mounted in it.

The CS finish was as for 1958, with the same options for the oil tank, its cover and the toolbox of blue for AJS, and white or red for Matchless, while the petrol tank was

An AJS 31CSR hitched to a Watsonian Monza sidecar and on test in 1967 when such outfits had become quite rare.

blue or red with various options of size. The CSR went back to black fork shrouds and rear unit covers, but kept the polished mudguards, chaincase and lower fork legs, the chrome-plated wheel rims and tank panels, and the options of AJS blue or Matchless red for the petrol and oil tanks, oil tank cover and toolbox.

Options for the CSR included the two-tone petrol tank, as for the tourers, and the de-luxe finish, except that the alloy mudguards remained. The Matchless twins could have the oil tank, its cover and the toolbox in white.

General options continued to be the steering damper, stop light, various handlebars and pillion foot-rests. For the tourers, there were accessories such as a rear carrier, crash bars and panniers, while these models could still have a saddle in place of the dualseat. All had alternative exhaust systems, so the siamesed type was listed for the tourers and the low-level twin version for the sports machines. The sports Matchless twins continued with the tubular silencer.

All this added up to a lengthy list of models with a great number of variations, despite the common forks, hubs and gearbox, plus just two basic engines and two frames. It was still a production nightmare, and steps were taken to rationalise it over the next three years.

The 1965 Matchless G15CSR which used the Norton 745 cc Atlas engine as well as forks and hubs from the same source.

The immediate move for 1960 was to drop the sports 500 twins, as customers for that style preferred the larger model. The de-luxe version went as well, as the touring rider was happy with coil ignition, and the better finishes remained options. The four larger twins remained, but for all models there were major changes.

The main change was to the front section of the frame, which gained duplex downtubes that ran back under the engine to the seat tube to complete the loop. The subframe was still bolted on, and there were two of these, one for the touring versions and the other for sports models. With the new frame went new dualseats, the touring seat being of the two-level type.

Other changes included a new cylinder head casting, on which the fin under the exhaust ports carried three smaller diagonal ones on its underside to assist the air flow. The headlamp shell was reduced in size, except in the case of the CS twins, but retained the same layout of the parts it carried. The finishes continued as they were, but where the two-tone tank was used on the tourers, there was the further option of leaving the other coloured items in black, so only the petrol tank was special.

Due to breakages in use, the firm began to make some of the crankshafts for the 646 cc engine in a nodular iron during 1960, and these became known as the 'noddy cranks'. More were used in 1961, for CS and CSR models, and in time all went over to the new material, which could withstand the power better without snapping.

The number of models dropped to four in 1961, as the larger CS twin went, to leave just the standard model in two sizes, and de-luxe and CSR versions of the larger. Of course, all were made as either marque. For all, there were shorter mudguards and larger tank badges, while the finishes were generally as before. What changed were the colour options, which were reversed so that the two-tone tanks were grey on top and blue below for AJS, and red top with white lower section for Matchless. To go with this, the single colour remained that used for the upper tank, so it became grey for AJS and red for Matchless.

That was the final year for the 498 cc twins, which had run their twelve-year course, and it was also

Matchless G15 for 1965 with Norton engine and other parts so only the frame and details were of AMC origin.

The 1966 AJS model 33 which used the Norton engine, forks and hubs while the gearbox had been common to AMC and Norton since 1957.

the finish for the 646 cc de-luxe machines. This left just the standard and CSR models in the two marques for 1962, which reflected the changing times and the beginning of declines in both the market and AMC's fortunes.

Changes were minimal, the CSR having an altered subframe, a new roll-on centre stand, and an ignition key to go with the coil ignition. The dynamo was dropped from the CSR, so it gained an alternator, but still retained its magneto. There were new die-cast alloy tank badges, much larger than in the past, which soon

became known as "knee-knockers'.

The finish for the standard models was black with chrome-plating and polishing as in the past. The one option was to colour all the painted parts, the AJS being in blue and the Matchless in red, but both with white mudguards and a grey seat. The CSR retained its alloy mudguards, with tanks and toolbox in blue or red, and had the same all-colour option with a grey seat, except that the mudguards remained in their natural alloy finish.

The option list was extended to include twin carburettors, of the same size as the standard one, and from May 1962 a headlight cowl for the CSR models only. This was a fibreglass moulding, with a fascia and small screen, that copied the style of the road-racing singles and was coloured to match the marque. The headlamp was mounted in it, together with a matching speedometer and rev-counter, as well as the usual ammeter and light switch, but used the normal AMC twin engine, stretched out some more, and a combination of standard, CS and CSR cycle parts. The two batches differed in minor details but both were in a bright, sporting style.

A further feature of 1962 was the use of quickly-forgotten model names. Thus, the 31 was the Swift, the G12 the Majestic, the 31CSR the Hurricane, and the G12CSR the

A 1966 Matchless G15CSR with its swept-back exhaust pipes, rear-sets, low handlebars and reversed gear pedal.

it was only offered for a few months. The rev-counter had been an option introduced for 1960 and was driven by a gearbox mounted at the left-hand end of the exhaust camshaft.

1962 was also the year that the 750 cc Matchless G15/45 was made in two batches of 100 each, all destined for North America. This model Monarch but, although AMC continued to use the names, few riders did, especially for the CSR machines.

There were quite a number of changes for 1963, and both models were given a new front hub with fewer fins and ball races in place of the taper-rollers. At the rear went a new subframe for the standard machine, and a revised rear fork to suit the adoption of standard Girling rear units without clevis end fittings.

Both models had their petrol tank capacity reduced to four gallons, thanks to the appearance of knee recesses in the sides. There were new dualseats of reduced width, and new silencers without tailpipes, so the lovely Matchless megaphone was no more, surely a bad marketing move. For the standard models alone, there was a rounder shape for the oil tank and toolbox, D-section mudguards and 18 in. wheels, which kept to the tyre widths of old.

The all-colour options were dropped, but otherwise the finish was essentially as it was for 1962. The option of blue or red tanks and toolbox, for AJS and Matchless respectively, continued for the standard models, while the CSR could have the tank sides chrome-plated.

There were further major changes for 1964, as the AMC twins adopted the Norton Roadholder front forks and Norton hubs and brakes. This meant a new frame, subframe and rear fork, and was all part of a standardisation programme to cut costs, as the group was in some financial trouble. The Norton wheels had full-width, light-alloy hubs, and the front carried a much needed 8 in. brake. While this was the usual single-leading-shoe type, it did mean that the Norton twin-leading-shoe brake from the later Commando could be fitted to further help in stopping the machine.

Both models changed to a 12 volt electric system, and the CSR adopted the rounder-shaped oil tank and toolbox, plus the 18 in. wheels of the standard model. The finish was much as before, but the AJS colour became polychromatic blue, and the CSR changed to chrome-plated mudguards and had the plated tank sides as standard.

Little altered for 1965, as the AMC twins were nearly at the end of their days, but the tank badges were altered to a diamond shape for AJS and a smaller letter 'M' for Matchless. The standard model colour options were no longer available, and the tank sides of the CSR were not plated. Then it was 1966 and no changes appeared the final year. The AMC twin had run its course and a great number of owners mourned its passing.

There was to be one final chapter in the story, but little of that was really AMC, aside from the Plumstead factory in which it took place.

With Norton engines

The last chapter of the AMC twin story is very heavily flavoured with Norton, for that AMC subsidiary supplied the engine, forks, hubs and brakes, while the gearbox was already common to the marques. This left just the frame and a few cycle components from the AMC parts bins, and the result was a hybrid that appealed to a limited circle of buyers.

The Norton engine was as old and as stretched as the AMC twin. It had grown to 745 cc to power the Atlas model, and it was this unit that went into the final AMC twins. The capacity came from 73 x 89 mm dimensions, and the compression ratio was 7.6:1. Engine construction began with a vertically-split crankcase, but both the iron block and alloy head were in one piece, the latter with integral rocker boxes.

The crankshaft was built-up and ran in two mains, while the camshaft went in the front of the crankcase, where it was driven by a combination of gears and chain. A second chain ran back to a rear-mounted magneto, an alternator went on the left-hand end of the crankshaft, and twin Amal

Last year for the AJS model 33 was 1967 and there were no changes for the end of the run.

Monoblocs supplied the mixture.

The rest of the machine was mainly as for the 646 cc models, with common frame, forks, wheels and gearbox. So were items such as the oil tank and toolbox, but there was a change of tyre section for the standard models. For them, the tyres were 3.50 in. front and 4.00 in. rear, but the 18 in. diameter remained. The CSR kept the tyres of the smaller twin, but had a very different appearance in the café racer mould. Thus, there were low bars, rearsets that folded up, a shortened brake pedal, reversed gear pedal, gaiters for the front forks, and a pair of magnificent, swept-back exhaust pipes.

There was also a rev-counter, driven from the right-hand end of the camshaft, with the gearbox attached to the timing cover. The head sat beside the speedometer on a bracket attached to the top of the forks, and the light switch went between them. This left just the ammeter in the headlamp shell, which was more in the form of the earlier separate type.

The finish for the standard models was as for the previous year's 646 cc CSR twins, so the petrol tank top, oil tank and toolbox were in polychromatic blue for AJS and red for Matchless. Mudguards and the tank sides were chrome-plated, while the AJS diamond and small 'M' badges were used on the tank. The CSR models had the same coloured and plated tanks and toolbox, but with polished alloy mudguards and chrome-plated headlamp shell, chainguard, wheel rims and instrument panel.

There were no changes for 1966,

The 1967 Matchless G15 before it became the Mk 2 the following year, but this was not to change much.

Sports AJS model 33CSR of 1967 with the same features as the Matchless, but not the charisma, for some reason.

but several for 1967, when a Matchless street scrambler, built for export, appeared as the G15CS. This had trail tyres, still on the 18 in. rims, and a small petrol tank, while being offered with lights, and also as the Norton N15CS with a simple change of tank badge! Either way, it was a handsome machine, with a candy apple red tank and chrome-plated mudguards.

The other models changed to 930 Concentric carburettors, and the tank capacity went back to the 1959 level. The CSR changed to 19 in. wheels, while keeping the same section, and the standard machine went to a 3.25 in. front while staying with the 18 in. rims. The finish of both remained as for 1966, plus the apparent option of a black Match-less tank for the standard model according to one brochure.

During the year, the two AJS models were dropped, so only Matchless machines were listed for 1968. There were still three of these, but the standard model became the G15 Mk2 and, together with the CSR, it changed to the capacitor ignition that had been used on the CS from the start.

The finish was unaltered, and in this way the three models continued into 1969 and the close of the factory. It was a sad end to a long-running firm that had been building motorcycles since the dawn of the industry, but for real AMC enthusiasts, the end had already come back in 1966.

Twins in competition

Both AJS and Matchless twins were used by factory riders in the ISDT in the early 1950s where the combination of road and trail work suited them. This began in the 1951 event where Hugh Viney used an AJS Model 20 suitably modified for the job, rather on the lines of the later CS machines. The next year both Viney and Manns were on twins with one from each marque and they repeated this in 1953 and 1954. For 1955 Manns and Usher both used the Matchless while for 1956 Manns was on an AJS enlarged to just over the 500 cc.

On the track, the AMC twins were involved in the Clubman's TT as early as 1950, when the engines were just polished a little and run on open pipes, as allowed then! In the following year, at the Manx Grand Prix, another twin appeared with an engine that had received considerably more attention and was installed in the racing AJS 7R cycle parts. It raised eyebrows, at this supposedly amateur event, and finished fourth.

A year later, in 1952, a similar works prototype won the Manx, which caused a controversy, but in October Matchless calmed the

The AJS twin used by Hugh Viney for the 1951 ISDT when he was a member of the British team that won the Trophy.

Paddy Driver on the Matchless G12CSR he shared with Joe Dunphy to win the 1963 1,000 kilometre production race at Oulton Park.

storm by announcing the G45 racing twin for 1953. This duly appeared, its engine being tuned with suitable camshafts and pistons, and alloy barrels with extra fins, but it was not too far from stock.

There were twin Amal GP carburettors to supply the mixture, a racing magneto and a rev-counter. The exhausts comprised pipes on each side with a megaphone for each, and these were tucked in very well. The rest of the machine was 7R, with duplex loop frame, Teledraulic front and pivoted rear forks, massive conical hubs with big drum brakes, racing Burman gearbox, one Imperial gallon oil tank, and racing seat

and controls, all finished off with a large petrol tank carrying the 'flying-M' transfers in silver. From then on, the G45 received detail improvements each year up to 1957, its final season. It was never very successful, but it provided variety on the racing scene and further choice to the rider.

A few CSR twins ran in production races in the 1960s, but this was far more the province of other marques. However, an AJS did take the 1960 Thruxton 500 mile event by a three lap margin, which showed that the machine had both speed and stamina - crankshaft and cam followers permitting!

AMC twins Specifications

All models have twin-cylinder, ohv engines driving a four-speed gearbox, a frame with pivoted-fork rear suspension and telescopic front forks.

AJS model	20	20CS	20CSR	30	30CS	30CSR
Matchless	G9	G9CS	G9CSR	G11	G11CS	G11CSR
years	1949-61	1959	1959	1956-58	1958	1958
bore mm	66	66	66	72	72	72
stroke mm	72.8	72.8	72.8	72.8	72.8	72.8
capacity cc	498	498	498	593	593	593
comp. ratio	7.0[1]	8.0	8.0	7.5	8.5	8.5
carb size	1	1	1	1[2]	1.06	1.06
ignition by	mag[3]	mag	mag	mag	mag	mag
top gear	5.25	5.8	5.25	5.25[4]	5.01	5.01
petrol - gall	4/3[5]	2 or 3	4.25	3.75	2 or 3	3.75
front tyre	3.25x19	3.00x21	3.25x19	3.25x19	3.50x19	3.25x19
rear tyre	3.50x19	4.00x19	3.50x19	3.50x19	4.00x19	3.50x19
front brake dia	7	7	7	7	7	7
rear brake dia	7	7	7	7	7	7
wheelbase in.	55.2	55.2	55.2	55.2	55.2	55.2

[1] - 1956-8.0
[2] - 1957-1.06
[3] - 1959-coil
[4] - 1957-5.01
[5] - 20-4.0, G9-3.0, 1954-both 3.75

AJS model	31	31CS	31CSR	33		33CSR
Matchless	G12	G12CS	G12CSR	G15[1]	G15CS	G15CSR
years	1959-66	1959-60	1959-66	1965-69[2]	1967-69	1965-69[2]
bore mm	72	72	72	73	73	73
stroke mm	79.3	79.3	79.3	89	89	89
capacity cc	646	646	646	745	745	745
comp. ratio	7.5	8.5	8.5	7.6	7.6	7.6
carb size	1.12	1.12	1.12	1.12[3]	930	1.12[3]
ignition by	coil	mag	mag	mag[4]	cap	mag[4]
top gear	4.79[5]	5.25	4.79[6]	4.51[7]	4.94	4.51[8]
petrol - gall	4.25[9]	2 or 3	4.25[9]	4.0[10]	2.25	4.0[10]
front tyre	3.25x19[11]	3.00x21	3.25x19[12]	3.50x18[13]	3.50x18	3.25x18[14]
rear tyre	3.50x19[11]	4.00x19	3.50x19[12]	4.00x18	4.00x18	3.50x18[14]
front brake dia	7[15]	7	7[15]	8	8	8
rear brake dia	7	7	7	7	7	7
wheelbase in.	55.2	55.2	55.2	56.5	55.4	56.5

[1] - 1968-G15Mk2
[2] - 33 and 33CSR to 1967
[3] - 1967-930
[4] - 1968-cap
[5] - 1963-4.59
[6] - 1960-4.51, 1966-4.32
[7] - 1968-4.72
[8] - 1966-4.22
[9] - 1963-4
[10] - 1967-4.25
[11] - 1963-18
[12] - 1964-18
[13] - 1967-3.25
[14] - 1967-19
[15] - 1964-8